Home-Study Experiments
to accompany

# Physics

## A Practical and Conceptual Approach

### SECOND EDITION

**JERRY D. WILSON**
Chairman, Department of Science and Mathematics
Lander College

Saunders Golden Sunburst Series

SAUNDERS COLLEGE PUBLISHING

Philadelphia
New York
Chicago
San Francisco
Montreal
Toronto
London
Sydney
Tokyo

Copyright   © 1989 by Saunders College Publishing, a division of Holt, Rinehart, and Winston.

All rights reserved.  No part of this publication may be reproduced or transmitt in any form or by any means, electronic or mechanical, including photocopy, recording or any information storage and retrieval system, without permission i writing from the publisher.

Requests for permissions to make copies of any part of the work should be maile to:  Permissions, Holt, Rinehart, and Winston, Inc., Orlando, Florida  32887.

Printed in the United States of America

Home Study Experiments to accompany PRACTICAL PHYSICS, 2/e

ISBN # 0-03-023769-6

901   066   987654321

# TABLE OF CONTENTS

**Introduction   Systems of Units   1**

   Experiment 1   Your Own System of Units   1

               2   Metric and British Unit Comparison   2

               3   Length Units (Including Anatomical)   4

               4   Volume and Mass-Weight Units   5

**Chapter 1   A Restless World: Motion, Force, and Newton's Laws   7**

   Experiment 1   Description of Motion (Speed)   7

               2   Accelerometer   7

               3   Inertia   8

               4   Falling Bodies   11

**Chapter 2   Work and Energy   13**

   Experiment 1   Work and Energy   13

               2   Power Expended in Climbing Stairs   14

               3   Conservation of Energy   16

**Chapter 3   Momentum   17**

   Experiment 1   Collisions   17

**Chapter 4   Projectile, Circular, and Planetary Motions   19**

   Experiment 1   Projectile Motion   19

               2   Ellipses -- the Orbits of Planets   21

**Chapter 5  Gravitation and Earth Satellites   23**

   Experiment 1   Acceleration Due to Gravity and Reaction Time   23

               2   g's of Force   25

               3   Relative Satellite Altitudes   23

               4   Apparent Weightlessness   26

**Chapter 6  Rotational Motion**  27

  Experiment 1  Rigid Body Rotation    27

           2  Cylinder Derby    28

           3  Conservation of Angular Momentum (Gyroscopic Action)  29

           4  Equilibrium and Stability   29

**Chapter 7  Atoms, Molecules, and Matter**  30

  Experiment 1  Chemical Bonding   30

           2  Polar Molecules   31

           3  Antimatter   31

**Chapter 8  Solids**  32

  Experiment 1  Crystalline Structure   32

           2  Lattice Structures   32

           3  Hexagonal Patterns   33

           4  Water Density and Temperature   33

**Chapter 9  Liquids**  34

  Experiment 1  Pressure    34

           2  Buoyancy   36

           3  Cartesian Diver or Bottle Imp    37

           4  Surface Tension   38

           5  Viscosity   39

**Chapter 10  Gases**  40

  Experiment 1  Atmospheric Effects   40

           2  Bernoulli Effects   41

**Chapter 11  Temperature and Heat**  42

  Experiment 1  Temperature Sense   42

           2  Molecular Theory   42

           3  Thermal Expansion (or Contraction)   43

           4  Maximum and Minimum Temperatures   43

           5  Radiation and Temperature   45

**Chapter 12   Heat Transfer and Change of Phase    46**

Experiment 1   Thermal Conductivity and Specific Heat    46

2   Heat Transfer and Relative Humidity    47

3   Dew Point Temperature    50

4   Freezing Point    50

**Chapter 13   Thermodynamics, Heat Engines, and Heat Pumps    52**

Experiment 1   The First Law of Thermodynamics    52

2   Efficiency of Transportation    53

**Chapter 14   Waves and Vibrations    54**

Experiment 1   Period and Frequency    54

2   Interference    54

3   Reflection and Standing Waves    55

**Chapter 15   Sound and Music    56**

Experiment 1   Resonance With a Finger Driving Force    56

2   Resonance Chimes    57

3   Beat Patterns    57

4   Air Columns    58

5   Musical Glasses    59

**Chapter 16   Electrostatics – Charges at Rest    60**

Experiment 1   Charging    60

2   Polar Molecules and Electrical Force    60

**Chapter 17   Electric Current – Charges in Motion    62**

Experiment 1   Lemon Battery    62

2   Electrical Energy    62

**Chapter 18   Magnetism    64**

Experiment 1   Making a Compass    64

2   Making an Electromagnet    64

3   Magnetic Force on Moving Electrical Charges    65

**Chapter 20  Light Waves**  66

Experiment 1  Fluorescence  66

2  Diffraction -- Single Slit and Gratings  66

3  Polarization  67

**Chapter 21  Reflection and Refraction**  68

Experiment 1  The Law of Reflection  68

2  Light Rays -- the Pinhole Camera  69

3  Right-Left Reversal and Nonreversal  70

4  Right-Left Reversal and Reversed Writing  70

5  Multiple Images  72

6  Refraction at a Liquid Boundary  73

7  Light Beam Refraction  73

8  Refraction and Depth  74

9  Conjuring Up a Coin  74

10  Refraction and Magnification  75

11  Waterdrop Lenses  75

12  Internal Reflection  76

13  Dispersion  77

**Chapter 22  Vision and Optical Instruments**  78

Experiment 1  Eyeglasses and Vision Defects  78

2  Color Addition  78

3  Color Subtraction  80

4  Color Dots  80

**Chapter 25  The Nucleus and Radioactivity**  82

Experiment 1  Atomic and Nuclear Dimensions  82

2  Black Box Experiment  82

**Chapter 26  Nuclear Energy: Fission and Fusion**  84

Experiment 1  Chain Reaction  84

# Preface

The physical principles you will study in your physics course are all around you. These principles and many practical applications will be described in the text of <u>Practical and Conceptual Physics</u> and your instructor will supplement these descriptions in class. However, it is also instructive and fun to investigate and apply some of the principles yourself.

This is the purpose of this <u>Home-Study Experiments</u> booklet. Some simple hands-on experiments are described that allow you to apply various principles. In some experiments, the items needed to perform them will be listed, while in others the one or two items needed are evident from the discussion. Generally, the materials are common and readily available -- you probably have or can easily obtain many of them and others may be inexpensively purchased. Your instructor may give you a helping hand in some instances. The topics of some chapters, for example, relativity, do not lend themselves to home experiments, but many do.

The experiments are designed to help you understand some of the topics studied in class, which should result in a better grade. That should be an enjoyable incentive, so enjoy.

                                      Jerry D. Wilson

                                      Lander College

# Introduction for the Student (Systems of Units)

Early measurement units were often referenced to parts of the human body, for example, the foot and the hand. These units varied from person to person, so there was a necessity for units to be standardized. As was learned in the Introduction of the text, the standard units of a measurement system are traditionally defined by a head of state or a government agency. Today, there are two major systems of units in use--our customary British system and the world-predominant metric system. The following experiments or exercises will help you understand the meanings of standard units and give a comparison of various systems of units.

**Experiment 1** Your Own System of Units

Try your hand at it. Devise a measurement system of your own (and give it a name). Select standards, names for these standards, and give the equivalents of your system's units with those of the SI system.

Name of System: _____


Standard Units (Give names and definitions. Don't forget abbreviations for units.)


Length: _____

_____


Mass: _____

_____


Time: _____

_____


Comparison of the _____ system and the SI system units

2    Introduction

Length:

        One _____ is equivalent

           to _____ meter.

One meter is equivalent to _____.

Mass:

        One _____ is equivalent

           to _____ kilogram.

Time:

        One _____ is equivalent

           to _____ second.

One second is equivalent to _____.

Give some possible submultiples and multiple units for your system. [For example, in the metric system we have 1000 grams in one kilogram. Also, 1000 kg is a metric tonne.]

**Experiment 2** Metric and British Unit Comparison

The metric system is here. Metric units are commonly found (along with British units) on many items. Probably most evident are the dual listings on food items. Go to the kitchen (or visit a grocery store) and find several of the following:

(a)    commercially packaged or canned products that list the contents in both ounces (weight) and grams (mass), and

(b)    liquid products that list both fluid ounces and milliliters. (Check the refrigerator.)

Record the unit comparisons below.

Introduction  3

Data:

(a) Weight-mass units

| Item | ounces (oz) | grams (g) |
|------|-------------|-----------|
| _____ | _____ | _____ |
| _____ | _____ | _____ |
| _____ | _____ | _____ |

(b) Volume units

| Item | fl. ounces | milliters (mL) |
|------|------------|----------------|
| _____ | _____ | _____ |
| _____ | _____ | _____ |
| _____ | _____ | _____ |

(Optional) From your data, find the conversion factors for the units in (a) and (b). For example, a length conversion factor is 1 inch = 2.54 cm or 1 cm = 0.39 in. [Note: although conversion factors are commonly written with an "equals" sign, e.g., 1 ft = 12 in., this means "equivalent to", in particular for weight-mass conversions, since they are not the same property. A ratio form is better, e.g., 12 in./ft, which is read, "12 inches per foot."] See the Extended View for the Introduction on conversion factors in the Appendix of the text.

# 4  Introduction

**Experiment 3**  Length Units (Including Anatomical)

Let's physically compare some length units.

<u>Items needed</u>:  some string, a pair of scissors, and a l-ft ruler that has dual calibration in inches and centimeters. [You could use a meterstick to measure metric units should you have one available, however these are not yet very common.]

(a) Cut a piece of string one meter in length and another piece one foot in length. Compare the string lengths.

   One meter is about _____ times one foot.

   One foot is about _____ times one meter.

(b) Cut a piece of string 30 cm in length and compare this with the 1 m and 1 ft lengths of string from part (a).

   30 cm is equivalent to _____ of a meter. 30 cm is about

   _____ times one foot, or about _____

   times one yard.

(c) Cut lengths of string equal to the units of hand, cubit, yard, and fathom. [Look up the lengths of these units in a dictionary if necessary.] Compare these to the previously cut lengths of string.

1 hand is about _____ inches or _____ cm. 1 cubit is about

_____ inches, or _____ cm, or _____ ft, or

_____ yd, or _____ m. 1 fathom is about _____

yd or _____ m.

Just for fun, compare the strings to the corresponding parts of your own anatomy. Also, cut a length of string equal to your anatomical foot. Compare this with a standard foot length and perhaps to the "foot" unit of another person. Comment on the practicality of the use of anatomical units.

**Experiment 4** Volume and Mass-Weight Units

In this experiment, we'll check out some standard units and see how good you are at estimating them.

Items needed: an ungraduated container, e.g., a large pitcher; a scales, e.g., bathroom; some sand or sugar; and water

[Note: the mass or weight of the container may be neglected in the experiment if it is not appreciable. This is usually the case when using a bathroom scales and a plastic container. However, if the empty container has an appreciable reading when placed on the scales, record its weight and work in terms of "net" quantities, i.e., subtract the weight of the container from the total weight readings.]

(a) Add sand or sugar to the container until it contains an estimated 1 kg mass of the substance. Then, weigh the container and record below.

(b) Add water to the (empty) container until it contains an estimated volume of 1 liter of water. Then, weigh the container and record below.

How do your estimates compare with the accepted values of the kilogram? Complete the Data Table. [1 kg has an equivalent weight of 2.2 lb and 1 liter of water has a mass of 1 kg.]

Data Table:

(a) Measured weight of sand or sugar _____

  Computed kilogram mass equivalent _____

(b) Measured weight of water _____

  Computed kilogram mass equivalent _____

Show calculations here and on next page:

Calculations:

Notes:

## Chapter 1  A Restless World: Motion, Force, and Newton's Laws

There's motion all around you. The description of motion was studied in Chapter 1, along with what causes motion, i.e., net or unbalanced forces, and the concept of inertia. The following experiments will help illustrate some of these principles.

**Experiment 1**  Description of Motion (Speed)

Objects travel at various speeds. We see this all the time. Estimate the average speeds of the following objects in <u>metric units</u>. The estimates may be made from experience, rather than from an actual experiment. However, you may wish to make some experimental observations to support and verify your experience.

(a) A jogger. [Hint: how many meters does a jogger travel in a few seconds? This information will allow you to give an estimate of the average speed in m/s.] Show any calculations.

Estimate: _____

(b) A marble or ball rolling across a table (in cm/s).

Estimate: _____

(c) A car passing a school, children present (in km/h).

Estimate: _____

(d) A turtle crossing a road.

Estimate: _____

**Experiment 2**  Accelerometer

In the text, an air-bubble level accelerometer was described. See if a marble or a ball can be used as an accelerometer.

8    Chapter 1

In a car, van, or bus, place a marble or a ball on a level floor and observe and record what happens in each of the following cases:

(a) The vehicle starts moving from rest.

(b) The vehicle is traveling at a constant velocity on a level roadway.

The vehicle is traveling at a constant velocity on a level roadway then:

(c) speeds up

(d) comes to a stop

(e) turns a corner

Does the marble or ball act as an accelerometer?  Explain in terms of each of the above observations.

**Experiment 3  Inertia**

Try the following cases and explain what is observed.

Items needed:  playing card or 3 x 5 index card, drinking glass, 3-4 dollars in quarters, heavy ball or object (or styrofoam cup full of marbles), and string (or tape)

(a) Place a playing card or index card on top of the glass and place a coin (quarter or half-dollar) on top of the card.  Flip the edge of the card with your finger so the card flies horizontally off the glass.  [Practice until you get the coin ending up in the glass.]

(b) Place a stack of 12-15 quarters on one end of a strip of paper (about 3 cm wide and 20 cm long) Give the other end of the paper strip a quick pull so as to remove the paper from beneath the stack. [Repeat if the quarters fall over.]

(c) Suspend a heavy ball or object as shown in the figure by a string (B) with a string (A) fixed to the bottom of the ball.
Note: if a heavy ball or object
is not available, R.D. Edge in
his <u>String and Sticky Tape Experiments</u>,
(University of South Carolina), suggests
using a styrofoam cup full of marbles
instead of a ball and tape instead of
strings. Equal cuts about two-thirds
through each tape are made so they will
break easily. (Are you prepared to pick
up marbles?) **Caution:** if a heavy object
is used in the following procedures, be
careful that it does not fall on a toe
or do other damage if string B breaks.

(i) Give string A a sharp, quick downward pull so the string breaks.

(ii) Repeat with a slow pull on string A until one of the strings breaks.

Record results and explanations on back of page.

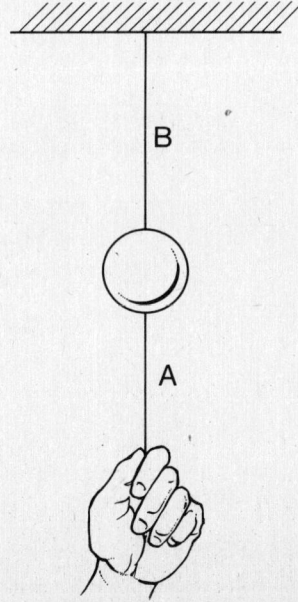

**Results and Explanations:**

**Experiment 4** Falling Objects

These cases aren't a Leaning Tower of Pisa experiment, but you should be able to explain them.

Items needed: a quarter, two pennies, and a sheet of paper

Drop the following pairs of objects simultaneously from equal heights. Observe what happens (which hits the floor first) and record and explain below.

(a)  two pennies

(b)  a penny and a quarter

(c)  a quarter and a sheet of paper (short sides horizontal)

(d)  a quarter and the sheet of paper crumpled into a loose ball

(e)  a quarter and the paper wadded into a tight ball (tie or tape together if necessary)

(f)  the tightly wadded paper ball and a penny.

Observations and explanations:

**Notes:**

## Chapter 2  Work and Energy

**Experiment 1** Work and Energy

As was learned in Chapter 2, work and energy are intimately related. To illustrate this, take a paper clip and bend it into a straight wire as best you can. Don't worry about the small curved regions of the original bends.

Hold the wire <u>near each end</u> and bend it back and forth as though you were trying to break it <u>near the center</u> of the wire (but don't). When you feel the wire weakening, hold the bending region to your cheek and record your observation and explanation in terms of work and energy below.

[If you continued flexing the wire it would break. Do you know why? If not, see the Question and Answer near the end of Chapter 8 for a sneak preview of the properties of solids.]


Observation and explanation:

## Chapter 2

**Experiment 2  Power Expended in Climbing Stairs**

Power is the time rate of doing work (or expending energy), or in equation form, $P = W/t$. Let's take a measure of your power output.

(a) Measure your average power output in walking up a flight of stairs. Record the measurements and show calculations in the data table below. [Hint: your work output will be approximately equal to the change in your gravitational potential energy (mgh), if you don't shuffle your feet too much. Note: work in SI units. Does it make any difference that your path is along the stairs rather than vertically upward?]

(b) Measure your <u>maximum</u> average power output by climbing the stairs as fast as you can. [**Caution:** don't over exert yourself and don't do this if you have a medical condition that might be aggravated.] Show data and calculations below. Convert your power output to horsepower to see how many horses you're worth.

<u>Data Table</u>

(a) Calculations

Mass _____
                                    (or weight)

Height _____

Time _____

Power _____

(b)   Calculations:                                    Time _____










                                                      Power _____

                                                      Horse-worth _____


<u>Notes</u>:

16     Chapter 2

**Experiment 3**  Conservation of Energy

Set up a pendulum with a bar or peg as shown in the figure. Describe and explain what is observed in terms of the conservation of mechanical energy at different portions of the pendulum's swing. [A pendulum suspended from the top of a doorway with a broom handle across the doorway works nicely.]

Have a measuring stick available to measure heights.

Observations and explanations:

        Initial height of pendulum bob _____

        Final height of pendulum bob _____
          (after hitting peg)


Observations and explanations:

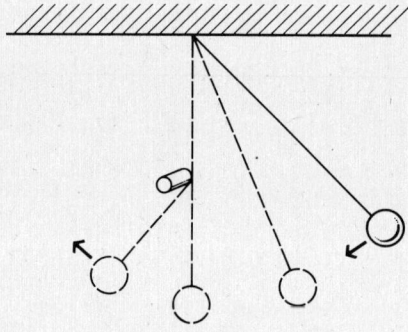

# Chapter 3  Momentum

**Experiment 1**  Collisions

A simple apparatus for studying collisions is shown in Fig. 3.10 in the text.  Set this up.

<u>Items needed</u>:  grooved ruler, and 6 marbles -- 5 of equal size (similar mass) and one larger (more massive).

Try the following cases:

(a) Place the four marbles of equal size (mass) in contact with each other near the center of the ruler.  Then, flick the other marble of equal size toward the stationary marbles.  Vary the number of stationary and flicked marbles.  (More or less than five marbles may be used.)  Also, try a one-on-one collision.

Observations and explanations:

(b) Try one-on-one collisions with the large marble and a small marble (marbles of different masses).  Flick the large marble towards the smaller one and vice versa.

Observations and explanations:

(c) Flick the large marble towards a stationary group of four or five marbles in contact. What is observed for this case? [Hint: in analyzing and explaining this situation, consider the initial collision and the transfer of momentum down the line. Then, go back to the big marble and think of what it does after the first collision. Is it still moving? Did it lose some momentum in the initial collision? Analyze the successive transfers of momentum down the line.]

Observations and explanation:

## Chapter 4    Projectile, Circular, and Planetary Motions

**Experiment 1**  Projectile Motion

Projectile motion may involve throwing or tossing an object. That's what will be done in this experiment. The path of a projectile depends on the angle of projection. The maximum angle is 90 degrees, or a projection straight upwards (a -90 degree projection would be directly downward). A horizontal projection is at an angle of zero degrees. In general, projection angles are between 0 and 90 degrees.

<u>Items needed</u>: a penny, two quarters (or two pennies and a quarter for the cheaper approach), a sheet of paper, and a pencil or marker

(a)  Hold a penny and a quarter on the open palm of one hand and throw them vertically upward, being careful that they leave your hand at the same time. Note whether they stay together to the top of the projection and whether they return together. (Let the coins hit the floor or a table.) Make several trials. Does projectile motion depend on mass?

Observations and explanation:

(b)  Place a coin on the edge of a table and hold a similar coin in the other hand at the same height. Simultaneously flick the coin on the table with a finger horizontally and drop the other coin. Make several trials. Do the coins hit the floor at the same time?

Observations and explanation:

(c) On a piece of paper, draw vertical and horizontal axes near one corner (horizontal axis along the long edge). Draw a 45 degree line and estimated 30 and 60 degree lines. Using the lines on the sheet as a guide, flip a penny with the thumb of one hand along the various angle lines and observe the ranges. Make several trials. (Try other estimated angles for fun and verification. Make sure to record these angles too.)

Observations and explanations: (Make drawings showing observations)

**Experiment 2** Ellipses

Let's get a feel for the shapes of ellipses and the locations of their foci by drawing a few.

<u>Items needed</u>: string, two thumb tacks, a pencil, and several sheets of paper

<u>Procedure</u>  With a loop of string and two thumb tacks (see Fig. 4.13 in text), draw the ellipses for the following cases:

(a) the foci three different distances apart (not maximum or minimum). Draw each of the ellipses on different sheets of paper.

(b) the special case when an ellipse is a circle.

(c) the special case when an ellipse is a straight line. [Note: don't force your pencil around the tacks. Their locations are points.]

Be neat. Your instructor may wish to see your drawings.

Give your personal description of ellipses below and how they vary with the distance between the foci. Also, considering that the Earth's orbit is elliptical but like most planets nearly circular, discuss how the Earth's climate might be if the Sun were at an orbital focus (singular of foci) nearer the orbital path.

**Notes:**

## Chapter 5  Gravitation and Earth Satellites

**Experiment 1**  Acceleration Due to Gravity and Reaction Time

We often think that our reactions are instantaneous. However, there is a finite time involved in the process of observing something, signals being sent to and from the brain, and the reaction taking place. This is called the <u>reaction time</u>. For example, if you suddenly notice a thrown ball coming toward you at a close distance, your reaction time is the time between when you see the ball and you start to move your hand to catch the ball or to protect yourself.

The reaction time is easily demonstrated and measured. This will be done in the experiment using falling objects. Also, you'll get an appreciation of the acceleration due to gravity and how fast things fall.

<u>Items needed</u>: a dollar bill, a ruler with centimeter graduations, and another person.

Note: you are going to measure your reaction <u>time</u> without a timer or watch.

(a)  Have your friend hold the dollar bill vertically your extended thumb and forefinger. The thumb and forefinger should be horizontal and parallel, and separated by several millimeters (see figure). The idea is form a "slot" through which the dollar bill will be dropped.

Then, with the short side of the bill just even with the top of your hand, have the friend drop the bill (without telling you when it will be released) and you try to catch the bill between your thumb and finger. If you're like most of us, you won't be able to do it. Explain why. (Dropping the bill yourself isn't cricket as you will tend to inadvertently "cheat".)

Explanation:

(b) Repeat the experiment with a ruler instead of a dollar bill. On catching the ruler, put the end of the dollar bill at the lower end of the ruler and see by how much you missed the bill.

To determine how fast your are, or your reaction time, repeat dropping the ruler and note the centimeter mark at the top of your finger. If you start with the zero end of the ruler downward, this is the length of the ruler that fell through your hand during your reaction time. Do this several times, recording the distances in the data table below, then find the average distance of fall.

Compute your reaction time. This can be done by using the equation

$$d = 1/2 g t^2 \quad \text{or} \quad t = (2d/g)^{1/2}$$

where the symbol $(\ )^{1/2}$ means square root, d is the distance of fall, g the acceleration due to gravity (9.8 m/s$^2$), and t the time. For example, suppose the ruler fell on the average a distance of 19 cm or 0.19 m. Then using your calculator:

$$t = (2d/g)^{1/2} = [2(0.19)/9.8]^{1/2} = 0.20 \text{ s}$$

(You could use 19 cm, but g = 980 cm/s$^2$ in this case.) Check the reaction time of your friend too.

Data: (show calculations)

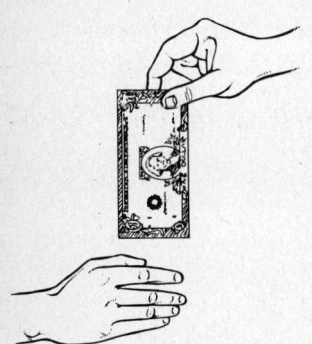

Reaction time: _____

Gravitation and Earth Satellites    25

**Experiment 2**  g's of Force

To get an idea of the g's of force acting on an airplane (and its pilot) in a circular loop as described in the text (see Fig. 5.13), swing a small pail of water (if you're daring) or a plastic milk bottle almost full of water (if you're not daring) in a vertical loop. While doing so, not the different g's of force that you must supply in different parts of the loop.

[Hint: start with a speed in excess of the minimum speed required to maintain the water in the pail (or have the cap on the plastic milk bottle). Otherwise, you may get wet. Why?] Caution: do not hit anything or anybody, including yourself when swinging the water container.

Observations and explanations:




**Experiment 3**  Relative Satellite Altitudes

Items needed: ruler, pencil, and paper

In drawings of satellites orbiting the Earth, they are usually exaggerated and not drawn to scale. Draw the Earth and the circular orbits of satellites at altitudes of (a) 600 km, and (b) 2000 km to scale. [You need to select the scale. For example, drawings of floor plans for a home may have a scale of 1 ft to the inch.] Be neat. Your instructor may wish to see your drawings.

Synchronous satellites, which have an orbital altitude of about 23,000 miles, have the same period of revolution as the Earth's rotation, and hence are relatively stationary over one location. Why? If this altitude were included on your drawing, how would it affect your scale?

## Experiment 4  Apparent Weightlessness

Items needed: several styrofoam drinking cups and water

Poke a small hole in the bottom of a styrofoam cup with a pencil or other pointed object. Fill the cup with water with your finger covering the hole. Remove your finger and check to see that a continuous, thin stream of water comes from the cup.  (Adjust the size of the hole or prepare a new cup if this is not the case.)

Refill the cup with a finger covering the hole.  Stand on a chair or some elevation Hold the cup out and allow a stream of water to flow from the hole (see Fig.  5.20 in the text). Then, drop the cup and note what happens to the stream of water while the cup is falling. This is best done outside or over a sink for obvious reasons.  (If not obvious, they will be after the first dropped cup.)

Since you have several cups, repeat several times.

Observations and explanation:

# Chapter 6  Rotational Motion

**Experiment 1**  Rigid Body Rotation

We have non-rigid bodies readily available that can quickly be converted into rigid bodies. Can you guess what they are?

Items needed: 6 or more eggs and a means of boiling them, a spoon, and timer or watch.
[Optional: plastic "eggs" used to package ladies' hose.]

Place 5 eggs in a pan of water and bring quickly to a boil. When the boiling starts, time the eggs for various intervals, e.g., 1 minute, 2 minutes, etc., and remove one egg with the spoon at the end of each interval. If an egg cracks during boiling, replace it with another egg. Be sure to identify the egg with the boiling time. (Be careful not to burn yourself when removing the eggs.) Leave the final egg in the water for an extended period to make certain that it is fully cooked or "hard" boiled. Allow the eggs to cool.

(a) To demonstrate the difference between rigid body rotation and non-rigid body rotation, use the raw egg and the hard boiled egg. Hold the raw egg on a flat table top and give it a quick spin with a sharp twist of the fingers and wrist so as to set the egg spinning. (Be careful not to give it a translational motion or you may have egg on the floor.) Note what occurs.

Try the same thing with the hard boiled egg. (If given sufficient torque, the hard boiled egg will "rise up" on end and spin like a top.)

[Note: the plastic "eggs" used to package women's hose may be used. Fill one with water for a non-rigid body and an empty one is a rigid body. Why?]

Observations and explanation:

28    Chapter 6

(b) The Earth is believed to have a liquid (outer) core (see Special Feature 14.1 in text), yet it is rigid enough overall to spin. Starting with the egg cooked for the least time, spin the egg and determine which egg as a function of cooking time is the first to rise up and spin like a top. Crack the eggs and inspect the yolks. What can you say by analogy about the Earth's "liquid" center?

Observations and conclusions:




**Experiment 2** Cylinder Derby

Items needed: several solid cylinders and hollow cylinders of different sizes, and an inclined plane. (Food cans make good solid cylinders, and the cans with the food removed and the ends cut out are hollow cylinders, as are napkin rings, etc. Keep in mind that this is a rigid body experiment. A can of juice or soda isn't a rigid body. Why?)

Release various pairs of one solid cylinder and one hollow cylinder simultaneously from rest at the top of an inclined plane. Note which cylinder reaches the bottom first. Also, try races with pairs of solid cylinders and pairs of hollow cylinders. Note the cans' relative masses (weights) and radii cans in these cases.

Results and explanations:

**Experiment 3** Conservation of Angular Momentum (Gyroscopic Action)

The next time you're riding a bicycle on a large, level surface, while sitting on the seat, lean your body slightly to one side (<u>not</u> too far for safety's sake) and note what happens. Try leaning the opposite way. Are there some angular momentum effects?

When riding along the edge of a paved road and the bike is in danger of slipping off, people have a tendency to lean toward the center of the road to prevent the bike from falling off. Is this the correct thing to do?

Observations and explanations:

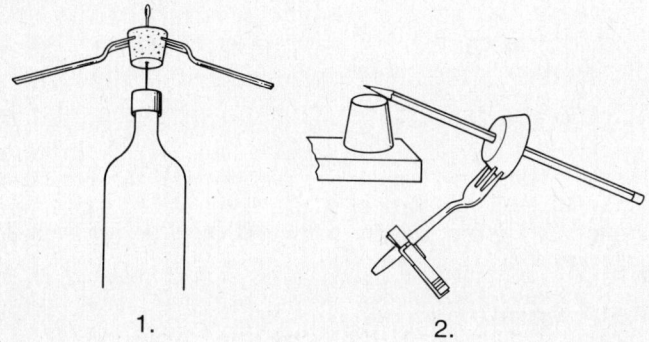

1.                      2.

**Experiment 4** Equilibrium and Stability

<u>Items needed</u>: a teaspoon, two forks, a wooden match, a drinking glass, a long needle, a couple corks or pieces of styrofoam, a bottle with cap, a pencil, and a clothes pin. (A variety of substitute items may be used.)

(a)    Construct the balancing act as shown in Fig. 6.26 in the text.

(b)    Construct the configurations shown in the figures below.

Indicate the location of the center of gravity relative to the base of support in each case [parts (a) and (b)]. Try some balancing acts of your own.

## Chapter 7  Atoms, Molecules, and Matter

You can even check out molecular bonding in simple experiments.

**Experiment 1**  Chemical Bonding

<u>Items needed</u>: small quantities of salt, sugar, and sand; metal spoon; iron skillet or frying pan; and heat source

Salt is an ionic compound and sugar is a covalent compound, as is sand (with a special type of structure).  Test for the following:

(a) Hardness.  See how difficult it is to crush a few grains of each substance with a spoon on a hard surface.  Record your observations below and rate the hardness.

(b) Heat stability.  Place a pinch of each substance in cast iron skillet or frying pan (or some other appropriate container), keeping them in separate piles.  [Your instructor may allow you to do this in a lab if this item is not available and the procedure inconvenient, e.g., in the dorm.]  Heat the bottom of the skillet slowly and evenly.  Record your observations and rate the heat stability below.  **Caution:** do not overheat or burn something (including yourself)

Explain your observations in terms of chemical bonds.

(a) Hardness

Sugar _____

Salt _____

Sand _____

[Does hardness depend on something in addition to chemical bonding?]

(b) Heat stability

Sugar _____

Salt _____

Sand _____

# Atoms, Molecules, and Matter

**Experiment 2** Polar Molecules

Perform the experiment shown in Fig. 7.10(b) in the text using a hard rubber comb. To charge the comb, rub it on a piece of fur or wool, or even through your dry (unoily) hair. Make sure the stream of water is as small as possible -- just enough to give a continuous stream.

[This experiment should only be done on a dry day. Do you know why? See Chapter 16 for the answer. Note what happens if the stream of water touches the comb and it gets wet.]

**Experiment 3** Antimatter

Obtain a small quantity of antimatter, and .... (just kidding).

# Chapter 8  Solids

**Experiment 1**  Crystalline Structure

<u>Items needed</u>: magnifying glass, small quantities of salt and sugar, and packet of artificial sweetener.

Sprinkle some salt and sugar on a dark surface. Examine and compare the grains of each in terms of the regularity of the shape of the grains. Just for fun, check out a few grains from a packet of artificial sweetener and compare to those of sugar.

Observations and explanations:




**Experiment 2**  Lattice Structures

<u>Items needed</u>: 20 pennies

See how many orderly patterns, or two-dimensional lattice structures you can make on a flat surface with the pennies touching each other ("close-packed"). Describe these patterns in terms of how many pennies touch a particular penny internal to the array, for example, "cubic" for four pennies in contact with a single penny. Determine which array is most compact.

Arrays and comments:

Solids    33

**Experiment 3** Hexagonal Patterns

[This may not be a good experiment if you live in a region with a warm climate.]

The next time it snows, catch some snowflakes on a piece of black cloth (velvet is good) or on a piece of dark colored glass or plastic. (Place the catching material outside or in a freezer beforehand so it is cold.)

Observe snowflake patterns and describe them.

**Experiment 4** Water Density and Temperature

Items needed: two thermometers, ice, and a tall glass of water

Put ice (crushed or cubes) into a glass of water so as to form a layer of ice on the top. Position the two thermometers in the glass so that one measures the temperature near the top surface of the water and the other measures the temperature at the bottom of the glass. The thermometers will come to equilibrium in a short time. Then, note and record any temperature change in either thermometer as long as a change occurs. Explain any changes.

Observations and explanations:

## Chapter 9 Liquids

**Experiment 1  Pressure**

<u>Items needed</u>: a brick and a ruler (with metric graduations)

The pressure a brick exerts on a surface depends on the brick's weight force and its contact surface area.  First, estimate the weight of the brick or weigh it if a scale is available (e.g., in bathroom). Record the weight in newtons. [1 lb = 4.45 N]

(a) Measure the lengths of the sides of the brick and compute the areas of the brick's three different flat surfaces (i.e., top or largest surface, side surface, and end surface) in square meters.

Then, compute the pressures in pascals the brick would exert on a table when laying on the different sides.  After doing this, hold the brick on your hand and see if you can detect the differences.  (The pressures experienced will not be the same in all cases. Why?)

(b) Estimate the areas of the long edge and a corner point of the brick and compute the pressures on a table if the brick were balanced (propped up) on these surfaces. Balance the brick on your hand on these surfaces. You may notice a difference.

Weight of brick _____ N

(a)                           Area                Pressure

  (top)        _____   _____

  (side)       _____   _____

  (end)        _____   _____

Calculations:

(b)                    Area           Pressure

    (edge)     _____   _____

    (corner)   _____   _____

Calculations:




Notes:

## Experiment 2  Buoyancy

<u>Items needed</u>: round or rectangular plastic bottle (about 20 cm tall) with cap, ruler, and wax crayon or felt-tip marker.

Draw a linear scale on the side of the bottle with divisions that indicate 0.1 of the bottle's height (or 0.1 of its volume).  Neglect the neck height or volume.  Then, fill a deep sink or tub with water to a depth that exceeds the height of the bottle.

(a) Add water to the bottle until it floats bottom down in the tub of water.  Taking the density of the water in the tub to be 1.0 g/cm$^3$, estimate the overall density of the bottle and its contents.  [Hint: how is the density related to the fraction of the submerged volume of the bottle?]

(b) Add more water to the bottle so it floats deeper (but not enough to sink it), and again estimate its overall density.

(c) Add as much water to the bottle as you can without sinking it.  What is the approximate overall density of the bottle in this case?

(d) Add some more water to the bottle so it sinks to the bottom of the tub.  What can you say about its overall density now?

Observations and comments:

**Experiment 3**  Cartesian Diver or Bottle Imp

<u>Items needed</u>:  clear plastic bottle with cap, a medicine dropper, and water.

Fill the plastic bottle full of water. Fill the dropper with water, but squeeze out enough so that it floats slightly below the surface (see the figure). [It is helpful to adjust the water in the dropper and float it in a sink first.] Then, screw the cap on the bottle.

Gently squeeze the bottle and watch the diver or "imp" take a dive. Note the volume of water in the glass portion of the medicine dropper when this is done. Explain why the imp goes down and comes back up. [Hint: think in terms of Pascal's and Archimedes' principles.

Observations and explanation:

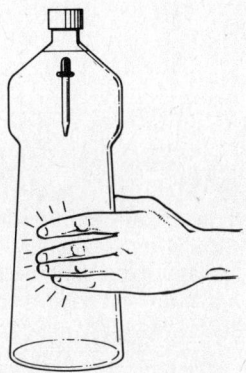

# Chapter 9

**Experiment 4** Surface Tension

<u>Items needed</u>: thin, double-edged razor blade (or light needle), detergent, large mouth bottle, and a piece of wire screen (e.g., window screen) larger than mouth of the bottle

(a) Try floating a razor blade (or needle) on water as shown in Fig. 9.9 in the text. Once you have it, add a few drops of detergent to the water and see what happens.

Observations and explanation:

(b) Fill the bottle with water and place the screen on top. (Make sure the screen is in good contact with the mouth of the bottle. Holding the palm on the screen over the bottle, invert the bottle. Slowly remove your hand so that you finally have two supporting fingers on the mouth of the bottle. [You might want to try this initially over a sink.] Explain why the water doesn't come through the screen.

**Experiment 5**  Viscosity

<u>Items needed</u>: test tube or small glass vial, small metal ball (a B-B or ball bearing) that will fit easily into the tube, wrist watch, and liquids found around the house (water, cooking oil, detergent, vinegar, etc.).

(a) Fill the tube almost full of water and place it against something so it leans a bit but is stable (see figure). Allow the ball to roll down the inside of the tube and time its descent to the bottom.

(b) Repeat the experiment with the other liquids with the tube at the same tilt.

(c) Categorize the liquids according to their relative viscosities.

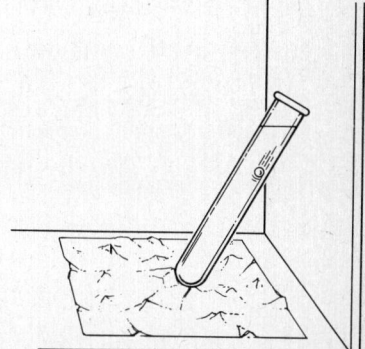

Data:

| Liquid | Descent time | Viscosity order |
|--------|-------------|-----------------|
| _____ | _____ | _____ (most viscous) |
| _____ | _____ | _____ |
| _____ | _____ | _____ |
| _____ | _____ | _____ |
| _____ | _____ | _____ (least viscous) |

## Chapter 10  Gases

It's easy to demonstrate the effects of two topics of this chapter -- the atmosphere and Bernoulli's principle.  Several are given for each in the following experiments.

**Experiment 1**  Atmospheric Effects

<u>Items needed</u>: a drinking glass, a 3 x 5 inch (or 7.6 mm x 12.7 mm in metric) index card, and a styrofoam or paper cup

(a)  Fill a glass with water the brim and place the card on top of it.  Holding the card, turn the glass over (see Fig. 10.20 in text).  Slowly remove your hand from the card and see what happens.  Can you turn the glass horizontally with the same result?  Try it and explain.  [You might try it first over a sink just in case.]

Observations and explanations:




(b)  While you have the glass handy, perform the experiment shown in Fig. 10.19 in the text.  A salt-water aquarium isn't necessary.  A sink and fresh water will do.

Observation and explanation:




(c)  Try the situation shown in Fig. 10.18 in the text.

Observation and explanation:    [Use the perfect gas law in your explanation.]

# Experiment 2  Bernoulli Effects

Items needed: sheet of paper, scissors, thin drinking straw, two Ping-Pong balls, string, tape, and vacuum cleaner with blower attachment

(a) For starters, perform the experiment shown in Fig. 10.15 in the text and explain.

(b) Cut the straw almost in two about midway along its length. Bend the straw so the two pieces are at a right angle with small uncut joining piece downward. Place one part of the straw in a glass of water and blow in the end of the other part. Note the water in the straw and explain. [If you put a very thin straw deep enough into the water and blow hard enough, you can get an atomizer effect.]

(c) Perform the experiment shown in Fig. 10.17 in the text. The stream of air is supplied by the blower attachment of a vacuum cleaner. [You may have to fashion a nozzle with a piece of paper so as to get a high-speed stream of air.]

(d) Suspend two Ping-Pong balls from equal lengths of string so they hang at the same height a couple of centimeters apart. (Use tape to attach the strings to the balls.) Then, blow between the balls and explain what happens. [This is why ships do not pass too close to each other. Also, you may have experienced a similar effect when driving on a freeway and an "eighteen-wheeler" passes you.]

## Chapter 11  Temperature and Heat

**Experiment 1**  Temperature Sense

Humans have a temperature sense and can feel relative hotness and coldness. Why then don't we just calibrate our hands for temperature readings (over the normal temperature range that would not cause injury) instead of constructing thermometers? The following procedures will give you an idea.

Fill three pans with water, one with the hottest water you can bear to put your hand in, one with luke warm water, and one with ice water (use ice).

(a)  Place both hands in the luke warm water and hold them there for about 15 seconds. Note the sensations in each hand.

(b)  Then, place your right hand in the hot water and your left hand in the ice water for about 15 seconds. Quickly dry your hands and plunge them both into the luke warm water again. Describe the sensations. What do you think about your temperature sense?




**Experiment 2**  Molecular Theory

<u>Items needed</u>:  balloon and tape measure

Blow up and tie a small rubber balloon. Measure its circumference with the tape measure and record. Place the balloon in the freezer compartment of a refrigerator or a deep freeze for an hour.

Measure the circumference of the balloon again and explain what happened in terms of molecular theory.

Observations and explanation:

**Experiment 3**  Thermal Expansion (or Contraction)

Most substances expand with increasing temperature. However, water contracts with increasing temperature over the range from 0°C to 4°C (cf. Special Feature 8.2 in the text). This is not the only substance whose dimensions decrease with temperature as will be seen.

Items needed: rubber band, small weight, ruler, and matches

Suspend a small weight on a rubber band so it is slightly stretched. Fix a ruler by the weight so that the height of the weight can be noted. Then, move a burning match (or cigarette lighter) up and down near the rubber band so that it is heated. [Heat thoroughly, but be careful not to burn the rubber band or yourself.] Observe what happens to the weight.

Observations and explanation:  [Speculate in terms of the rubber's molecular makeup, cf. Chapter 7.]

**Experiment 4**  Maximum and Minimum Temperatures

The daily maximum and minimum (high and low) temperatures are commonly given on weather reports. These temperatures may be read directly from continuously recording thermographs, but they can also be read from special maximum and minimum thermometers. Let's investigate one of these that you have no doubt used before.

## Chapter 11

(a) Obtain a clinical thermometer and measure your body temperature. Record below and compute your temperature in degrees Celsius and kelvins.

Calculations:

Body temperature

_____ °F

_____ °C

_____ K

[Note: $T_C = 5(T_F - 32)/9$ and $T_K = T_C + 273$, cf. Chapter 11 Extended View in the text Appendix.]

(b) While you were doing your calculations, the thermometer will have cooled and be at room temperature. Yet, if you read the thermometer again, it still registers your body temperature. Is the clinical thermometer a "maximum" thermometer? Examine the thermometer and explain why the mercury column does not fall as the thermometer cools. [How do you reset the thermometer?] Do a little research and report how a low temperature is measured with a "minimum" thermometer.

**Experiment 5**  Radiation and Temperature

In Special Feature 11.3 in the text, a list of the various colors of visible radiation emitted by a hot body and the corresponding approximate temperatures is given. Using this information, turn an electric stove burner on high and note its temperature increase by the change in color. (Do this in a darkened room.)

Letting the burner cool or using another burner, start timing from when the burner is turned on and record the times of the various color changes or temperatures. Draw a graph of temperature versus time below and plot your data.

Data and graph:




Questions: (answer on separate sheet of paper)

1. Does the burner heat evenly? If not, can you explain why?

2. What is the approximate temperature of a glowing cigarette or cigar? Explain your answer.

3. Is the temperature of a gas flame uniform? If not, give the approximate temperatures of the regions. [Your instructor may wish to set up a Bunsen burner in class and let you take some data.]

# Chapter 12  Heat Transfer and Change of Phase

**Experiment 1**  Thermal Conductivity and Specific Heat

<u>Items needed</u>: two styrofoam drink cups, two paper drink (unwaxed), water, matches, and a candle

(a) Light the candle and place one of the styrofoam cups slightly above the flame. Note how easily the bottom of the cup melts. Repeat with a paper cup. [Caution: this should be done near a sink so any flame may be quickly extinguished. Also, be careful to not burn yourself.]

(b) Fill the other paper cup with about one-quarter to one-half full of water and hold it slightly above the candle flame. Is a hole burned in the cup? (The bottom of the cup may turn black from the residue from the candle.) Explain why the water makes a difference. See how hot you can get the water. Can you make it boil? (Move the cup around to promote even heating and mixing.)

Observations and explanation:

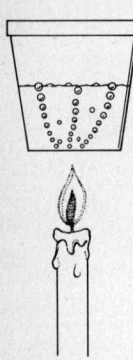

(c) Do you think a styrofoam cup with water will spring a leak when held over the flame? Try it and see.

Observations and explanation:

**Experiment 2** Heat transfer and Relative Humidity

Items needed: two thermometers, black felt-tip marker or pen, piece of cloth, and thread

(a) Blacken one of the thermometer bulbs with the marker. (Blacken with paint if ink will not adhere to the glass.) Set the thermometers out in the sun for a while and take the temperature readings.

Observations and explanation:




(b) Remove the blackening from the thermometer, and tie a cloth sleeve around the bulb. The sleeve should extend 4-6 cm below the bulb. Wet the cloth with water and place the tail of the sleeve in a container of water so the cloth around the bulb will remain wet by wicking (capillary) action. Do not place the bulb directly in the water. [Hint: the thermometer may be fixed on a paper milk carton with some water in the bottom. Make a hole in the carton and run the sleeve through to the water.

Set this and the other thermometer out of the way on an inside counter overnight. Note and explain any differences in the readings of the thermometers the next day.

Observations and explanation:




You can leave the setup for several days and take the readings of the thermometers each day. Call them the dry-bulb and wet-bulb temperatures. (Add water if necessary.) The apparatus you've constructed is called a psychrometer, which is used to measure relative humidity. Knowing the air-temperature (dry-bulb reading) and the depression of the wet-bulb (dry-bulb reading minus wet-bulb reading), you can find the relative humidity, as well as the dew point, from the following psychometric tables. Check your dew point result with Experiment 3.

Table 1

**Table 2**

## Experiment 3  Dew Point Temperature

<u>Items needed</u>:  glass of water, ice, and thermometer

A general approximation of the dew point temperature can be obtained by putting ice into a glass of water and stirring with a thermometer.  Notice the temperature at which condensation just starts to appear on the outside of the glass.  This is approximately the dew point temperature of the air.

Try this on a couple of different days when the relative humidity is high and low (measured in Experiment 2 or from the weather report) and compare how much the air temperature would have to be lowered to have 100 percent relative humidity, i.e., when the air temperature is equal to the dew point.  What would you expect?

Observations and explanations:

## Experiment 4  Freezing Point

<u>Items needed</u>: ice cubes, thread, and salt

(a)  Take two ice cubes and hold them together gently.  Notice that they don't stick together.  Then, push them together firmly as hard as you can. [Stack the cubes on a counter top and push down on the top cube using a towel so you don't slip.] Do they stick together this time?

Observations and explanation:

(b) Float an ice cube in a glass of water. Carefully lay one end of a 20-30 cm length of thread across the top of the cube and sprinkle some salt on the ice along the thread. Wait a minute or two, then lift the other (unsalted) end of the thread.

Observation and explanation

[Can you make a string of ice cube beads?]

Notes:

# Chapter 13  Thermodynamics, Heat Engines, and Heat Pumps

**Experiment 1**   The First Law of Thermodynamics

<u>Items needed</u>:  thermometer, electric blender, and paper clip

(a)   Fill the blender about half full of water and let sit until the water is at room temperature. Measure the temperature with the thermometer. <u>Remove</u> the thermometer and cover and turn on the blender.  Stir the water vigorously on the highest blender setting for about a minute.  Stop and measure the temperature of the water again. Stir the water more and again measure its temperature.

Explain what is observed thermodynamically.  Would the effect be different if you used some other liquid?  [Note: a pan of water and a hand egg beater may be used, but the energy will come from another source in this case.]

Observations and explanation:

(b)   Take a paper clip and bend it into a straight wire as best you can.  Don't worry about the small curved regions of the original bends.  Hold the wire <u>near each end</u> and bend it back and forth as though you were trying to break it <u>near the center</u> of the wire (but don't).  When you feel the wire weakening, hold the bending region of the wire to your cheek.  Explain this effect thermodynamically.  [This is a repeat of Expt. 2, Chapter 2, where it was considered in terms of work and energy.]

Observation and explanation:

**Experiment 2** Efficiency of Transportation

Although not exclusively thermal efficiency, heat engines are involved. This experiment will help you appreciate the overall efficiency of part of your activities.

Obtain a map of the local community. Mark the locations of your residence, the post office, bank, pizza parlor, stores, etc., that you visit most frequently. Identify about 10 or more locations in all. Measure the distances from your residence to each of these locations and the distances between each. Make a sketch of these locations and label the distances below.

What type of transportation do you usually use to travel from one location to another? Discuss the efficiency in terms of fuel consumption, time, and cost of the transportation system you use. Would it be difficult to improve the efficiency?

Data and conclusions:

## Chapter 14  Waves and Vibrations

**Experiment 1**  Period and Frequency

<u>Items needed</u>: simple pendulum and timer (watch or clock)

(a)   Set the pendulum oscillating in a small arc. Determine the average period of oscillation by timing several oscillations, e.g., 4-5, and dividing the total time by the number of oscillations. Compute the frequency of oscillation.

Data and calculations:

(b)   Change the length of the pendulum and repeat part (a). Does this affect the period and frequency?

(c)   Change the mass of the bob and repeat part (a). Does this affect the period and the frequency?

**Experiment 2**  Interference

<u>Item needed</u>: toy Slinky

Have another person help you stretch the spring on a long table or the floor. Then, at the same time, both of you produce a pulse or disturbance in the spring. [Practice doing this. Pull back a group of coils and release for a longitudinal pulse and give a sideways shake for a transverse pulse.] Carefully observe and describe what happens when the pulses meet and interfere.

Also, observe the reflection of a single longitudinal pulse. Note the speed of the wave pulse in the spring. Stretch or put more tension in the spring and observe the pulse speed again. Is there a difference? [Try some other tensions if you like.]

Observations and explanations:

**Experiment 3** Reflection and Standing Waves

(a) Obtain a length of rope (15-20 ft, e.g., clothes line) and tie one end to a rigid support such as a door knob. Stretching the rope loosely, give your free end a quick upward shake to produce a pulse. Observe the pulse's reflection from the fixed end. Note any phase change on reflection. Repeat with a downward shake to produce a downward pulse.

Observations and explanations:

(b) Shake the end of the rope up and down at different rates and see how many different standing waves you can produce. Note how many harmonics you can produce. Explain why the amplitude at the antinode positions are much larger than the amplitude movement of your hand.

Observations and explanations:

## Chapter 15  Sound and Music

**Experiment 1**  Resonance With a Finger Driving Force

<u>Items needed</u>: crystal glass or brandy snifter (A thin wine glass made of regular glass will sometimes substitute, but crystal glass is better.)

(a)  Wet your finger with water, and holding the base of the glass firmly on a counter or table with the other hand, <u>carefully</u> run your wet finger lightly around the rim of the glass.  A slight pressure may have to be applied.  When done properly, you will hear a sound or the glass will "sing".

Observation and explanation:

(b)  Put some water in the glass so it is about one-quarter full.  Repeat the above procedure.  Note any difference in the sound and explain.

(c)  Add more water to the glass and repeat.  Do this for 2 or 3 levels of water.  When the glass is almost full, note the vibrations in the water around the sides of the glass.  Explain.

## Experiment 2  Resonance Chimes

<u>Items needed</u>:  teaspoon, tablespoon, fork, and 3-ft length of string.

Make loops at each end of the string that will allow your forefingers to go through. Then, tie the teaspoon to the middle of the string (with one loop so it can easily be removed). Put your forefingers through the loops and into your ears. Then, lean over and strike the spoon against something solid, e.g., a chair. Do you hear something different?

Repeat with the other spoon and fork. Note any variations.

Observations and explanations:


## Experiment 3  Beat Patterns

<u>Items needed</u>:  several combs (with same and different tooth spacings).

Take two combs and with their flat sides together, run them back and forth over each other observing through the comb teeth. What do you see? Investigate with combs having same tooth spacing and combs with different tooth spacings.

The patterns seen with combs of unequal tooth spacings are called Moire patterns. They are analogous to sound beat patterns since they are produced by a difference in the tooth "frequencies" of the combs. See if you can relate the number of "beats" seen with the difference in the number of teeth per length of the combs.

Observations and comments:

**Experiment 4**  Air Columns

Items needed: bottle (e.g., beverage), several plastic drinking straws, and pair of scissors

(a) Hold the side of the mouth of the bottle against or near your lip and blow into it. When blowing at the proper angle you will hear a sound. Pour some water into the bottle and repeat. Do this for several levels of water.

Observations and explanation:

(b) Take a plastic straw and hold one end flat. Cut a "V" in the end and then chew the end of the straw. Chew with the back teeth so as to crumple and mesh together flat the "V" ends. Chew the straw slightly forward of the "V". When ends are crumbled, make sure the ends of the "V" are not stuck together and place this end in your mouth so the lips touch the beginning of the round part of the straw. Blow into the straw. [If no sound is produced, make sure the "V" ends are flat, and/or chew some more on the straw, or try another straw. With a little practice, you will get the effect.] Why is it necessary to chew the straw to get a sound?

Once you have it, while blowing into the straw and producing sound, use the scissors to cut off about an inch of the straw, then another inch, then another inch, and so on, until you get close to your lips. Be careful not to cut your lip or nose. [If you make 6 cuts on the straw, you will hear a familiar sequence. Adjust the cut lengths for this.]

Observations and explanations:

**Experiment 5**  Musical Glasses

<u>Items needed</u>:  6 to 8 identical drinking glasses (relatively thin-walled), and a wooden kitchen spoon

Tap gently on the side of a glass near the top with the wooden spoon.  Note the tone.  Try this with a couple other glasses.  Are the tones similar?  Why?

Add water to the glasses so they all have different levels.  (You may keep one glass empty if you like.)  Tap on the glasses with the spoon and note the tones.  Are they different?  Why?

Adjust the water levels in the glasses so as to give the tones of the musical scale.  [How many glasses will you need to make the tones of one octave?]  See if you can play a simple tune on your musical glasses.

Observations and explanations:

## Chapter 16  Electrostatics -- Charges at Rest

**Experiment 1**  Charging

Try charging various objects as described in the text, e.g., a balloon, hard rubber comb, yourself, etc.  Experiment with other objects you think might be charged.  How about a plastic pen?

Using charged objects, try to observe evidence of the electrical force, for example, will a charged comb repel a charged balloon?  Explain how or the method by which the various objects were charged.  [Recall that this experiment should not be attempted on a humid day, unless you want a lot of negative (not in the electrical sense) results.  Why?]

Experimental results and explanations:

**Experiment 2**  Polar Molecules and Electrical Force

If you haven't tried it yet, do Experiment 2 in Chapter 7.  (Do it again even if you have done it.  It's fun.)  Explain this effect in terms of the electrical principles learned in Chapter 16.  [This experiment should not be done on a humid day.  Why?]

Explanation:

**Notes:**

## Chapter 17  Electric Current -- Charges in Motion

**Experiment 1**  Lemon Battery

<u>Items needed</u>:  lemon, copper wire, and a paper clip

A simple battery cell can be made from a lemon (or lime).  Straighten out a paper clip and stick this wire and a similar piece of copper wire about half way into a lemon and about 3-4 cm apart (see figure).

Touch the wires to your tongue.  Do you feel a slight tingling sensation?  Explain why.  What is the electrolyte of this battery?

Observations and explanations:

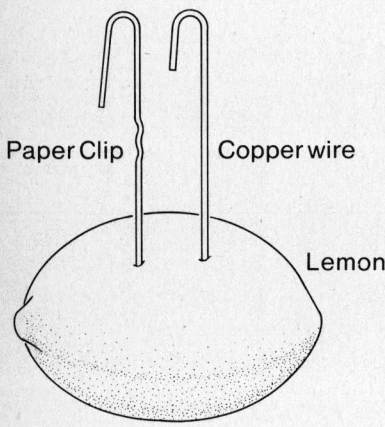

**Experiment 2**  Appliance Electricity Costs

(a)  Check the wattage ratings on a half dozen commonly used appliances, e.g., TV set, stereo, hair dryer, etc.  Estimate how long each is used daily and calculate the electrical energy each uses per day in kilowatt-hours.  [Recall $P = W/t$ or $W = P \times t$ or energy = power x time (kW x h = kWh).]

Find the cost per kilowatt-hour in your area from an old electric bill and compute the cost of using each appliance per month (30 days).  [If you live in a dorm, your instructor can give you the kWh charge to use.]

Show data and calculations on next page.

(b) Go to the business office of your school and request the monthly electric and/or heating costs for the institution for several months, preferably for all of the months of the academic year that the dorms are fully populated. Compute the average energy cost per person. [You'll need to know the student enrollment, and don't forget to include the faculty and staff. They use electrical energy too.]

Give some ways the energy consumption at your institution could be reduced. [You might try to get these put into practice and avoid a tuition increase.]

Data and results: (a) and (b)

## Chapter 18  Magnetism

**Experiment 1**  Making a Compass

<u>Items needed</u>: a strong magnet, a needle, and a cork or piece of styrofoam

Magnetize a needle by stroking (in one direction) on or with a magnet.  Run the needle through a cork or piece of styrofoam so the needle sticks out both sides.  Float this in a pan of water and see if you can determine the direction of magnetic north.  Give the direction in relationship to two well-known landmarks or buildings.  [How will you be able to tell north from south?]

Observations and results:


**Experiment 2**  Making an Electromagnet

<u>Items needed</u>:  an iron bolt or several nails, insulated wire, and a battery (dry cell <u>only</u>).
    **DO NOT** use an auto battery which has the chemical capability to give too much current and cause bad burns.

Wrap a number of turns of wire around the iron bolt or a bundle of nails (see figure).  Connect the ends of the wires (insulation removed) to the battery terminals.  [Disconnect one of the wires when not using the electromagnet so as not to run the battery down.] If the electromagnet doesn't appear to work, you may have to connect more than one battery together.  Which would you use, a series or parallel connection?  Also, use a large number of coils (more than shown in the figure).

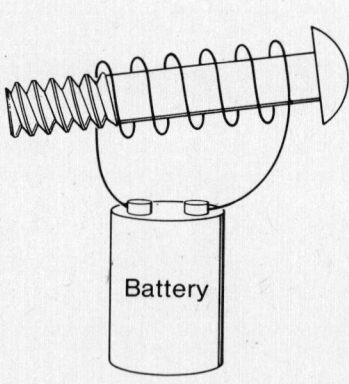

Try to pick up small pieces of metal, e.g., paper clips, etc. See how the strength of the electromagnet varies with the number of coils around the metal core. If you have a compass available, check out the magnetic field of the electromagnet with and without the metal core.

Observations and results:










**Experiment 3** Magnetic Force on Moving Electrical Charges

Items needed: a magnet and a black-and-white TV set.

The screen of a television set is activated by an electron beam in the TV tube that regularly scans the screen back and forth in horizontal lines. Hold a magnet to the glass front of a black-and-white TV picture. Move the magnet around and see what happens.

**DO NOT** do this with a color TV set. There are metal screens or masks in the tubes of these sets which could become magnetized. You'd then have a continuous effect of what you'll observe on a black-and-white set.

Observations and explanation:

## Chapter 20  Light Waves

**Experiment 1**  Fluorescence

Obtain some brightly colored laundry detergent boxes and/or breakfast cereal boxes and compare the brightness of the colors under (a) an incandescent lamp, (b) a fluorescent lamp, and (c) in sunlight. [If you have a black light available, try it too.]

Observation and explanation:

**Experiment 2**  Diffraction -- Single Slit and Grating

Items needed: index card, razor blade, candle, matches, and a feather

(a) Cut a 2-3 inch slit in the index card. Standing about 3 feet from the lighted candle in a darkened room, hold the card close to the eye and look at the candle flame through the slit. [The slit may be made wider or narrower by bending the card.] Do you see a "halo" around the flame? Is there color?

Observations and explanation:

(b) Hold the feather to your eye and look at the candle flame in the darkened room. Turn the feather to get the best observed pattern. Are colors observed? What are these? What type of grating is this?

**Experiment 3**  Polarization

Items needed:  two "lenses" from a pair of polarizing sunglasses or two Polaroid sheets (or two pairs of sunglasses if you don't want to remove the lenses)

(a) With the lenses from the sunglasses, show how one lens can be used as a polarizer and the other as an analyzer.  If you didn't know, could you determine the transmission axes direction in this manner?  Could you use a lens to determine if another pair of sunglasses were polarizing? Explain.

(b) View different regions of the sky on a clear sunny day through one of the lenses. Rotate the lens while viewing.  Is skylight polarized?  Is there some relationship to the region of maximum polarization and the position of the Sun?  [Check this out systematically.]

(c) View the "glare" reflected from a road or water or shining surface, e.g., a car, on a sunny day through a lens of polarizing sunglasses as you would normally look through it.  Then, rotate the lens in front of your eye noting and explaining any differences.

(d) Using the lenses, perform the procedure described in Question 23 in Chapter 20 of the text and experimentally verify your answer to the Question.

# Chapter 21  Reflection and Refraction

The topics of some chapters don't lend themselves to simple experiments. However, there are plenty of reflection and refraction experiments and they are fun. For most of the following experiments on reflection, you'll need one or two plane rectangular mirrors. It would be worth the cost to purchase a couple inexpensive mirrors if you don't have them available. You could always use them after doing the experiments and you'll have a greater appreciation of their reflective properties. In any case, be sure and try Experiment 4(c) below. The author really had fun with it.

**Experiment 1**  The Law of Reflection

<u>Items needed</u>: comb with even tooth spacings, plane mirror, and sheet of white paper

Fix the sheet of paper on the cover of a book (cut to size) and hold the comb along the edge of the book so that only the teeth protrude (see figure). Position or slant the book in sunlight so that several long parallel beams and shadows are seen.

Then, hold the mirror vertically on edge on the paper so the beams strike it. First hold the mirror at a right angle to the beams and note the effect. [It may be helpful to tilt the mirror slightly forward.] Place the mirror at other angles to the incoming beams. Could you verify the law of reflection? Explain how.

Observations and explanation:

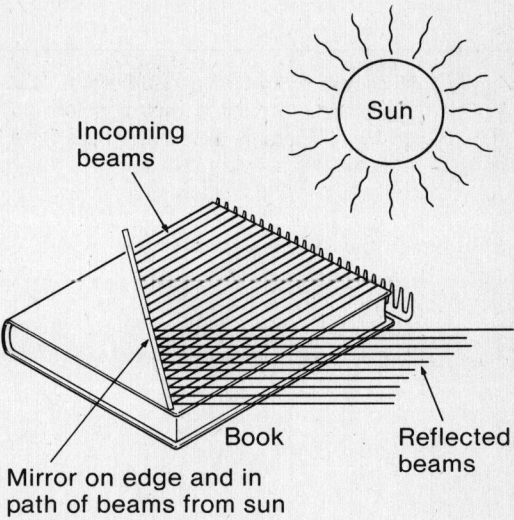

**Experiment 2**  Light Rays -- the Pinhole Camera

<u>Items needed</u>:  a rectangular or cylindrical cardboard box, e.g., a cylindrical cereal box, candle, piece of waxed paper (about 3 inches square), piece of black paper (about 2 inches square), tape, and a needle or pin

Cut a hole about an inch square in one end of the box and tape the black paper over this hole (see figure).  Punch a small hole in the center of the paper with a needle or pin.  In the other end of the box, cut a hole about two inches square and tape the waxed paper over this hole.

Then, in a darkened room, hold the end of the box with the pinhole 3-4 inches away from a burning candle.  Move the box back and forth until a sharp image is observed on the waxed paper.  Note the orientation of the image and explain.  [Hint: recall light travels in straight lines.  Trace some rays from different parts of the candle.]

This arrangement is sometimes called a pinhole camera.  Could it be used as a camera?  Explain how.  Would there be any magnification of an image?

Observations and explanations:

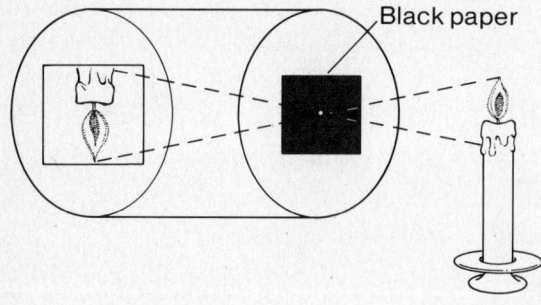

Black paper

**Experiment 3**  Right-Left Reversal and Nonreversal

Items needed: two plane mirrors and tape

(a) Look at yourself in a plane mirror. Is your image reversed? [Wink your eye or pull on your ear and see what the image does.] Hold the writing on this page up to the mirror. Try the face of a clock.

(b) Set up two mirrors with their edges touching and at a right angle or 90 degrees to each other (see figure). [Tape along the edge of the backs of the mirrors so they will stand freely.] Look into the common edge of the mirrors so that half of your image is in one mirror and half in the other. Try winking and pulling your ear now. (This is how people see you.) Hold the printed page and the clock in front of the mirrors too. Try performing some simple tasks like parting and combing or brushing your hair, shaving (if you do so), etc., and see how you do.

Observations and comments:

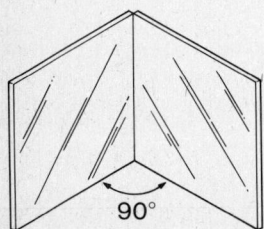

**Experiment 4**  Right-Left Reversal and Reversed Writing

Items needed: pen, sheet of paper, sheet of carbon paper, and two plane mirrors

(a) Hold this page up to a single mirror (or vice versa) and try reading looking into the mirror. Then, look at the image of the first mirror with a second mirror. [Look into the second mirror holding it so you can see the first.]

Observations and explanation:

(b) Place a sheet of carbon paper, <u>carbon side up</u>, under a sheet of plain paper. Write something on the plain paper, e.g., your name. Then, look at the carbon copy of the writing on the bottom of the plain paper. Try looking at the writing in a mirror with the edge of the mirror on the paper at the right side of the writing and then the edge of the mirror at the top of the writing. Also, view with the edge of the mirror at the left and bottom.

Observations and explanations:

(c) Sometimes you can "fool" the right-left reversal or it will fool you. For example, look at the following word in a mirror with the edge of the mirror at the right side of the word.

### HOE

Now, try looking at the word with the edge of the mirror above the word. Try another with the mirror in both positions:

### WEOM

Also, try looking in the mirror at the following word with the mirror held at the sides and above and below the word:

### WOW

Observations and explanations: [Make up several nonreversing words of your own. Which letters can you use? Consider and label only sideway nonreversals and both sideway and top-and-bottom nonreversals.]

**Experiment 5  Multiple Images**

(a) Set two plane mirrors upright so their edges touch and form a right (90 degree) angle. [See figure in Experiment 3.] Tape along the backs of the mirrors so they will stand freely. Place an object between the mirrors, for example, a coin or a marble, on line with their common edge. If you use a coin, how much money do you see? (Make you feel richer?) [Tilting the mirrors slightly may help to see more images.] See if you can increase the number of images by varying the angle between the mirrors.

Observations and explanation:

(b) Set up two mirrors upright and parallel with the reflecting surfaces facing each other. [Prop up the mirrors with something so they will stand.] Place an object, e.g., a coin or marble or something taller, between the mirrors. Look over one of the mirrors into the other and see how many images you can see. How about in the other mirror? [Tilt the mirrors to accommodate your eye if necessary.]

Place a brightly colored object between the mirrors. Can you notice that some images are not as bright as others? [Hint: not 100 percent of the light is reflected on each reflection.]

[As a variation, and if you want to see your "clone" images, stand in front of a mirror and hold another mirror by your head facing the other mirror. Look at the image of the second mirror in the first. Adjust the hand-held mirror slightly if necessary. What do you see?

Observations and explanation:

# Reflection and Refraction

**Experiment 6**  Refraction at a Liquid Boundary

<u>Items needed</u>: rubbing (isopropyl) alcohol, water, and a small glass

Pour some alcohol into a glass. Then add a small amount of water. Both the alcohol and water are clear, colorless liquids. Describe and explain what you see. Question: can you devise a test to determine if any two clear, colorless liquids are miscible (capable of being evenly mixed)?

Observations and explanation:




**Experiment 7**  Light Beam Refraction

<u>Items needed</u>: piece of dark cardboard (darken one side with a pencil or pen if necessary), clear drinking glass, water, and a few drops of milk

Make a small hole in the cardboard and add a few drops of milk to the water in the glass so as to slightly cloud the water. Then, place the glass in direct sunlight and hold the cardboard in front of the glass (dark side toward the glass) so a beam of sunlight will shine through the hole into the glass (see figure). A flashlight in a darkened room may also be used.

Hold the cardboard so that the hole is just below the water level and observe the direction of the beam in the water. Then, raise the cardboard until the beam strikes the surface of the water. Observe the direction of the beam. Experiment with the beam striking the surface at different angles. [What is the purpose of the milk in the water?]

Observations and explanations:

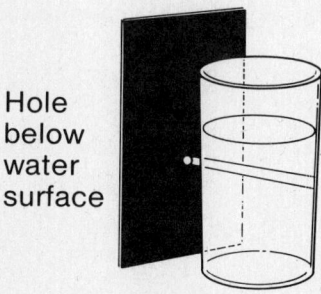

Hole below water surface

**Experiment 8**  Refraction and Depth

<u>Items needed</u>:  clear drinking glass (at least 5 inches tall), water, coin, and piece of aluminum foil

Crumple a piece of aluminum foil and smooth it out again. Set a glass of water on the foil and look straight down into the water. Does the bottom of the glass appear closer than the foil outside of the glass?

Drop a coin into the water and look straight down at it. Does it appear closer than the foil outside the glass? While looking at the coin, put your finger on the outside of the glass at the height the coin appears to be. What fraction is this of the height of the water?

Observations and explanations:  [Hint: we think of light as traveling in straight line, but it is refracted at a boundary. Draw a sketch showing this with the length of an extended straight ray in the water the same length as the refracted one.]

**Experiment 9**  Conjuring Up a Coin

<u>Items needed</u>:  a pan, a coin, water, and some help.

Place a coin in the pan near the side and position yourself so you can just not see the coin for the side of the pan (see figure). [Come close, then back away so you just cannot see the coin. Rest your chin on something fixed, e.g., your arm on the table, so your head does not move.] Then, have a friend add water to the pan and see what happens.

Observation and explanation:

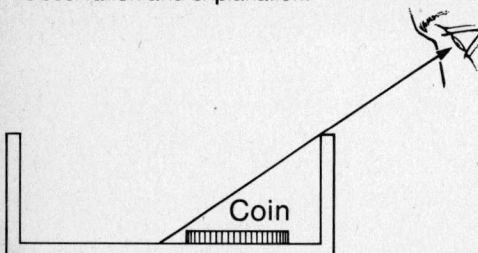

**Experiment 10** Refraction and Magnification

Items needed: clear, thin-walled glass (a stemmed water or wine glass works well), water, and a pencil.

(a) Fill the glass about three quarters full of water. Place a pencil (or a pen) in the glass and hold it so it stands vertically in the center of the glass. Observe the pencil through the side of the glass at eye-level. Is there a difference above and below the water surface?

Then, let the pencil rest on the side of the glass (don't hold it), and observe through the glass again. Do you see another effect?

Observations and explanations:

(b) Remove the pencil and fill the glass to the brim with water. Hold your hand over the rim and turn the glass horizontally. (Make sure there is not an air bubble in the glass.) Looking through the water-filled glass, view some lettering or printing. Explain your observations.

**Experiment 11** Waterdrop Lenses

Items needed: aluminum foil, glass plate (or slide), and water

(a) With a sharp object, e.g., a pencil, make a hole in the aluminum foil (about 1/16-1/8 inch in diameter). Place a drop of water on the hole and look through the drop at some print. Is the printing magnified? If so, estimate the magnification. How many convex surfaces does the drop "lens" have?

(b) Place a drop of water on a glass plate. Look at some print through the drop. Is there any magnification in the case? If so, estimate how much. How many convex surfaces does the drop have in this case?

**Experiment 12** Internal Reflection

<u>Items needed</u>: clear drinking glass, candle, water, and a plane mirror

Set the candle on a table a foot or two from the edge and light it. Place a glass of water nearby in an elevated position at the edge of the table. (Sit the glass on a stack of several books.) Then, kneel down by the table and look at the under surface of the water. Adjust your angle of view (or the candle) until you see an image of the candle flame. Would a mirror held horizontally produce a similar image? Try it and see.

Observations and explanation:

# Reflection and Refraction

**Experiment 13** Dispersion

<u>Items needed</u>: clear drinking glass, pan, plane mirror, several pieces of white paper, and water

(a) Place a glass full of water on the edge of a window ledge in bright morning or evening sunlight with several pieces of paper on the floor extending from the wall below the glass. With the right position of the glass and paper, you should see a spectrum. Explain.

(b) Place a mirror in a pan of water as shown in the figure. Position the pan in the bright sunlight and use a piece of white paper as a screen as shown (a foot or two distance from the mirror). Adjust the screen to find a spectrum. Note the order of the colors and explain why the spectrum of colors is seen.

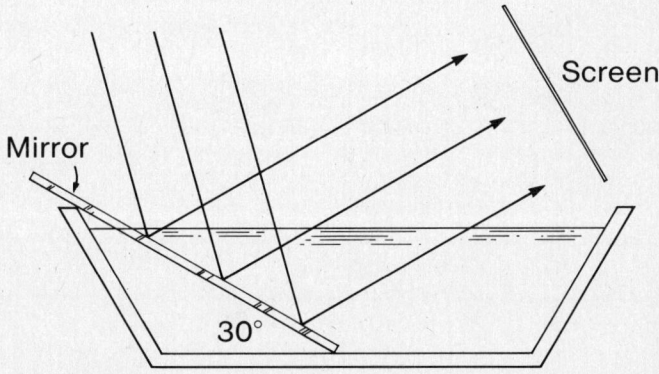

## Chapter 22  Vision and Optical Instruments

**Experiment 1**  Eyeglasses and Vision Defects

<u>Items needed</u>:  several pairs of eyeglasses and a flashlight

Determine what types of lenses are in several pairs of eyeglasses by shining a beam from a flashlight through a lens and trying to focus the light on a wall or floor.  (It may be helpful to darken the room.) Determine whether a lens is converging or diverging and speculate whether the person wears glasses for nearsightedness, farsightedness, or astigmatism.

Observations and explanations:




**Experiment 2**  Color Addition

<u>Items needed</u>: color paints, markers, or crayons; cardboard; pencil; scissors; and some string

(a)  Paint central circles of two complementary colors e.g., blue and yellow, on each side of a cardboard circle and attach strings as shown in the figure.  Twirl the strings between your fingers and thumbs so the cardboard rotates when the strings are pulled.  [Another method is to fix the cardboard on the eraser end of a pencil and rotate the pencil between the palms of the hands.] What color(s) would you expect to see?  (The shades of the colors need to be carefully selected to get the desired result.) Try several disks and different complementary colors.

Observations and explanations:

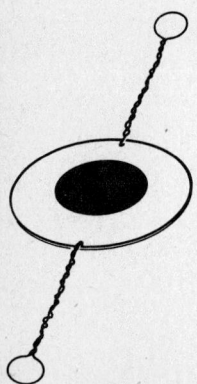

(b) Construct a color twirler as illustrated in the figure. Color the segments of the disk as shown and twirl the disk by winding up the string and pulling and relaxing on the ends so the disk rotates. Observe the combination of colors. Try other disks with different fractions of primary colors and half disks or disks with alternate radial segments of only two primary colors.

Observations and explanations:

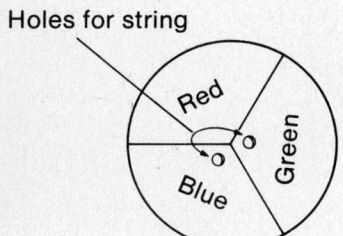

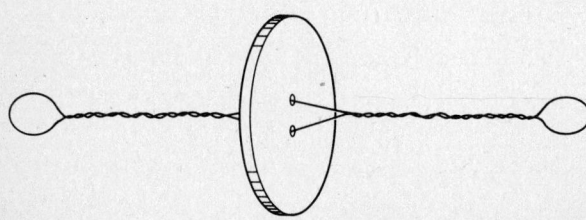

Chapter 22

**Experiment 3** Color Subtraction

Items needed: a clear glass, a set of food colorings, and different colored pieces of transparent plastic

(a) Experiment with color subtraction by mixing drops of food coloring in a glass. Do the primary colors of red, green, and blue produce white?

Observations and explanations:

(b) View objects of various colors through pieces of transparent plastic of different colors. Explain your observations in terms of color subtraction.

**Experiment 4** Color Dots

Use a magnifying glass to look closely at a color TV picture or a colored page in a magazine. Are these examples of color addition or subtraction?

**Notes:**

# Chapter 25  The Nucleus and Radioactivity

**Experiment 1**  Atomic and Nuclear Dimensions

On a sheet of paper, draw the structure of a typical atom to scale. [This is similar to drawing the floor plan of a house to scale. For example, the scale for a house plan might be 1 meter per cm i.e., 1 cm on the drawing represents 1 meter.] Don't forget to write the scale you choose on the drawing. [Hint: a jagged symbol in a line is used to indicate and imply scale distances that will not fit on a drawing.]

Conclusions: (attach drawing)

**Experiment 2**  Black Box Experiment

In many instances in science, things must be inferred from indirect observations or evidence. For example, we cannot actually see the nuclear structure. It is as though the atomic nucleus were inside a black box into which we cannot see and must indirectly try to find out what's inside.

To illustrate this idea, with another student, obtain two boxes with lids about the size of a shoe box. [Shoe boxes work fine.] Without letting the other person see, each of you place some regularly shaped objects (3 or 4 of them, e.g., a cube, cylinder, etc.) and seal the box with tape.

Then, exchange boxes and using your wits, try to determine the shapes of the objects in the box (without opening it, of course). Check your results and reasoning when you think you've gotten all of them.

Assumed object shapes and reasons for assumptions.

Object 1:

Object 2:



Object 3:



Object 4:



Notes:

## Chapter 26  Nuclear Energy: Fission and Fusion

**Experiment 1**  Chain Reaction

<u>Item needed</u>: set of dominoes

Set up a chain reaction using dominoes.  Place one domino on end, then less than a domino length in front of this two dominoes, then three dominoes, and so on, in parallel rows so a domino in one row is behind the space between two dominoes forward to it.  [The dominoes in a row should be separated less than the width of a domino.] Consider yourself to be a neutron and knock over the first domino to start the chain reaction.

Can you set up an arrangement whereby there is a single intermediate "neutron" domino for each domino in a row starting with the initial row with two dominoes?  Try angling the dominos.

Observations and explanations:

**Notes:**